ASQ's Pocket Guide Series

Writing Effective Audit Reports

By Denise Robitaille

QUALITY PRESS
Milwaukee, Wisconsin

Writing Effective Audit Reports
Denise Robitaille

American Society for Quality, Quality Press, Milwaukee 53203

Publisher's Cataloging-in-Publication Data

Names: Robitaille, Denise E., author.
Title: Writing effective audit reports / Denise Robitaille.
Series: ASQ's Pocket Guide Series
Description: Milwaukee, WI: Quality Press, 2026.
Identifiers: LCCN: 2026936826 | ISBN: 9781636942353 (paperback) | 9781636942360 (pdf) | 9781636942377 (epub)
Subjects: LCSH Auditors' reports. | Auditing. | Business report writing. | Business communication. | Business writing. | BISAC BUSINESS & ECONOMICS / Auditing | BUSINESS & ECONOMICS / Business Communication / General | BUSINESS & ECONOMICS / Business Writing
Classification: LCC HF5719 .R63 2026 | DDC 808.066651021—dc23

ASQ advances individual, organizational, and community excellence worldwide through learning, quality improvement, and knowledge exchange.

Bookstores, wholesalers, schools, libraries, businesses, and organizations: Quality Press books are available at quantity discounts for bulk purchases for business, trade, or educational uses. For more information, please contact Quality Press at 800-248-1946 or books@asq.org.

To place orders or browse the selection of all Quality Press titles, visit our website at: http://www.asq.org/quality-press.

30 29 28 27 26 LS 7 6 5 4 3 2 1

QUALITY PRESS

Quality Press
600 N. Plankinton Ave.
Milwaukee, WI 53203-2914
Email: books@asq.org
Excellence Through Quality™

Contents

Preface

Over the years, I have helped numerous organizations implement an ISO 9001 compliant quality management system. Many of these companies have been small, with total staffing numbering anywhere from 20 to 150 individuals. Internal auditor training is a module in the overall project.

It has been my practice to encourage top management to enlist individuals at all levels of the organization for the internal audit program. Additionally, to foster the independence of the auditors, I generally have them select individuals from multiple departments and functions throughout the organization. What usually ends up happening is that people from the manufacturing and warehouse areas conduct audits of the administrative-type processes (often thought of as "front office" functions), such as purchasing, sales, and design. The customer service reps, purchasers, and engineers audit production, inspection, receiving, and shipping. The remaining processes are meted out accordingly.

Top managers often raised their eyebrows and expressed profound doubts as to the capability of some personnel to conduct audits. The thought that these same individuals would also write their own audit reports seemed incredible—after all, many of the people chosen had only a high school diploma or a GED equivalent. Some had even less education. For a few, English was their second language. And yet, with only one or two rare exceptions (over a total of more than 20 years), they have all succeeded in asking probing questions, citing perceptive findings, and writing cohesive and meaningful audit reports.

I remember asking the president of one company, "Have you read their audit reports?" Several hours later, he caught up with me over a break. He expressed his surprise, enthusiasm, and appreciation over what he had read.

Writing a good audit report isn't hard. It requires diligence and commitment. It obliges the auditors to think about the risk and importance of what they are writing, as well as the consequences for their coworkers. Invariably, auditors are undeterred and honest when they generate their reports. They are engaged in the conversation about contributing to the improvement of the place where they work.

Frequently, I've had new auditors tell me how much they learned about the company they work for. "I didn't know we did this!" "That's why we have to do those checks in the morning." "Wow, I just talked to the engineers about our new design." They are often excited.

But, their real contribution comes through in the audit reports. Their reports make a difference.

Nothing in this book is hard. It simply requires the willingness to be engaged in the organization's management system and contribute to the improvement process.

—Denise Robitaille

Introduction

What is an audit report, and why does it matter? Consider: Why can't an auditor just sit down at the end of an audit, do a debrief, chat with the auditees, report any findings, let them ask their questions, and then just leave? No fuss, no muss, no extra stuff. What extra benefit is derived from the report?

After all, if the auditee is paying attention, they should have taken note of what was said and, presumably, taken the appropriate action. It is their organization. They should decide if they're going to heed the words of an auditor or not.

Except that the audit report *does* matter. The correctness and completeness of the information that is derived from the audit, and the conclusion reached after careful deliberation of what has been objectively observed, all matter. The actions that are taken as a result of the audit findings matter.

Without the audit report, the information that you have originates from an unsubstantiated source. Basically, it's a variation on the "he said, she said" scenario, and there's nothing backing up either party's recollections or assertions. Facts and details could have been missed, overlooked, or misinterpreted. Without the audit report, you have findings that aren't supported by documented evidence. You're left without any documentation that represents the integrity and value of the audit process. You could walk out of that closing meeting feeling as though someone has just wasted your time.

An audit report is important. It elevates the auditing process from a casual visit or fault-finding foray. It is the ultimate output of the auditing process. It results from the professional practices of a trained auditor. Competent and qualified auditors interview individuals based on documented standards and procedures using auditing tools and well-developed plans. They utilize sampling methodologies and risk-based approaches to determine the criticality and the appropriate amount of information to gather. They are persistent in following audit trails. They augment their work by assessing the complexity and adequacy of support processes such as document configuration, monitoring, communication between ancillary processes, and environmental controls. They deliberate over what they've read, heard, and observed, and they arrive at responsible conclusions as to the status of a process—or an entire organization. All this is accomplished through activities rooted in objective evidence that is accurately distilled and dutifully reported.

The report provides confidence that the information is correct and that it is presented in a format that the auditee can understand. It allows for follow-up actions to proceed unencumbered by the nagging doubt that the status reported during the closing meeting was just based on opinion and conjecture.

It provides justification, particularly to top management, for the expenditure of money and resources to address any findings that may have arisen. It enables them to calculate the return on investment (ROI) for the time and money the company spends fixing a problem or mitigating a potentially catastrophic risk that has been identified during the audit.

It's an ultimate assertion to the organization that they can have confidence in the evidence gathered and the conclusions that were reached.

Chapter 1
What Is an Audit Report?

The audit report is the product of the audit. Without it, the audit is incomplete. It is the culmination of all the auditing activities that preceded it. It provides invaluable information that benefits two unique stakeholders: the audit client and the auditee.

As defined in ISO 19011, the auditee client is the "organization or person requesting an audit." The auditee is the "organization as a whole or parts thereof being audited." For each organization, the actions that ensue from the audit report are at least partially dependent on the purpose of the audit. Subsequent chapters further discuss the scope and purpose. For now, it's fair to say that, regardless of purpose, the general intent of an audit is to address problems and initiate actions to improve.

An audit is a process. And, like all other processes, it follows the model of inputs, activities, and outputs. The audit report is the ultimate output of the audit process.

Applying the process approach model (Figure 1.1), we can follow the flow. The inputs include the defined scope of the audit, its purpose, the criteria to be used, and the evidence that is accumulated during the course of the audit—which will be used to substantiate the audit conclusion. Thus, as with any other process, the outputs should correspond to the inputs.

Inputs	Activities	Outputs
• Audit scope • Audit criteria • Evidence accumulated	• Interviewing • Observing • Assessing document • Reviewing records	• Evidence • Findings • Audit report

Figure 1.1 The process approach model.

Let's say the scope of the audit is production. It will include all processes such as production planning, multiple operations, and verification and/or inspections and tests. It could also include, as appropriate, ancillary processes such as material traceability, control of related documentation, packaging, labeling, and environmental controls. During the audit, the ensuing activities will occur:

- Interviewing personnel
- Assessing documentation
- Observing practices
- Reviewing records

The evidence that is evaluated and accumulated will run the gamut from production schedules to job travelers to inspection reports to material certificates and a variety of other bits of evidence. It could also include observation of the actual product, labeling practices, environmental conditions, and individuals performing various operations. All these are followed by aggregation of and deliberation on the result of these activities, which result in the eventual final output: the audit report.

The audit report is predicated on the premise that the inputs and the outputs should match. (See Figure 1.2.)

Inputs	Activities	Outputs
• Manufacturing procedure and/or work instructions • Industry and/or international standard • Production schedule • Evidence accumulated	• Interview operators and inspectors • Review drawings, job routers, machine setup instructions, sampling plans, inspection criteria • Observe operations being performed • Note any applicable environmental controls	• Evidence in the form of travelers, inspection reports, material traceability records, etc. • Observations of evidence against the input standards • Audit report

	• Review records generated: SPC charts Material certs In-process inspections Other records	

Figure 1.2 Comparing inputs and outputs.

To take the example one step further, let's look at two unique requirements. (See Figures 1.3a and 1.3b.)

Inputs	Activities	Outputs
Inspection sampling plan WI-076 Rev B requires that for a lot of 1,000 pieces, a random sample of 80 pieces must be pulled and inspected.	Interviewed the machine operator and observed the selection of the sample size followed by the inspection process. Reviewed the in-process inspection record.	Observed parts pulled randomly from four separate bins. Checked records of the last three inspections for work order #78-5766 and found inspection lot size correct in all instances.

Figure 1.3a Example requirements.

Inputs	Activities	Outputs
Procedure 76-01 Rev B requires that all suppliers be verified prior to placing an order for products or outsourced services.	Interviewed the purchasing manager. Reviewed the status of three suppliers in the company's enterprise resource planning (ERP) system. Reviewed three recent purchase orders.	Noted that two of the purchase orders (one for components and one for third-party testing) had been issued to suppliers whose status couldn't be verified in the ERP database.

Figure 1.3b Example requirements.

In Figure 1.3a, the example shows the inputs match the outputs. The activity conforms to the requirements. In Figure 1.3b, the example shows, the output does not conform to the documented requirement.

These examples demonstrate the objective evidence that is the foundation for the audit report. It's important to note that each audit finding has equal validity. They both clearly and objectively articulate the status of a particular process at a specific point in time—namely, the time of the audit. The organization needs both bits of information. They need to know what's working and what isn't.

It's important to reiterate that at this point in the audit, the findings are simply statements of fact. According to ISO 19011, they are based on "results of the evaluation of the collected evidence against audit criteria." ISO 19011 further goes on to state that "audit findings can lead to the identification of risks, opportunities for improvement or recording good practices." It is the auditor's decision that will ultimately determine if a finding should be raised as a nonconformance. This will be based on consideration of possible additional evidence that would mitigate the risk or that would suggest that alternative controls have been implemented.

The conclusion and decision to raise a finding of nonconformity as opposed to an opportunity for improvement is particularly relevant in those instances where the findings are graded. This is the case in conformity assessments, such as those conducted by certification bodies against an international standard or when a scoring checklist is used for a supplier audit.

Bear in mind that the importance of understanding the interrelation of processes will affect the deliberation process. If ancillary or support processes are robust and deliberately deployed to mitigate risk, the auditor could ultimately arrive at the conclusion that a process is well controlled despite nonconformance to a documented requirement. This illustrates the application of one of the principles of auditing found in *ISO 19011—Due professional care*: the application of diligence and judgement in auditing. The audit report will reflect the auditor's deliberation and reasoning in arriving at the audit conclusion.

Notwithstanding the due diligence and judgment of auditors, there are some circumstances wherein the auditor has limited choice. For many regulatory audits and some conformity assessments, any audit finding that shows nonconformance to a documented requirement must be reported. Chapter 5 discusses findings of nonconformity in more detail.

One of the quality management principles found in ISO 9000 that manifests itself in good audit practices relates to the process approach. It describes the mandate for interrelated processes to "function as a coherent system." Here's where some of the ancillary or support processes come into play in an audit. If the processes all work well together, the integrity and effectiveness of the entire system can be observed. These are the observable practices that might change a finding of nonconformance, requiring corrective action to an observation of perceived risk that may be articulated as an opportunity for improvement that should be assessed by the auditee and appropriately addressed.

If the audit scope is the machining of a component in conformance with documented specifications, the evidence should substantiate that the output of the process (the product) conforms to the specifications. The evidence should exceed a cursory verification that the product meets the physical criteria. There must also be confirmation that it was manufactured by individuals who have the requisite competence, are performing tasks in the prescribed method and sequence, in a suitable environment, using machinery that is well maintained, as well as raw material that has the necessary certifications, and that the product was subsequently verified using instruments that have been calibrated or otherwise verified. The myriad requirements inherent in the previous sentence represent various clauses in many management standards. Therefore, rigorous observation of one manufacturing operation could yield evidence of conformity to a host of unique requirements.

In each instance, the criticality of these factors is reliant on the nature of the product, the risks engendered, and the customers' expectations. In some cases, a factor may be nonapplicable. The effective interrelation of these requirements must be verified to determine if the process is adequately controlled. The audit report should include a narrative that reflects consideration of these factors.

The audit report contains valuable information about the status of an organization or a specifically audited process or group of processes. The audit report can, and should, shed light on the health of a company's management system and the extent to which the management system is a benefit to the organization. Its value should exceed its ubiquitous and long-standing reputation as an unpleasant but necessary litany of gaps, failures, problems, and violations.

An audit report is intended to impart specific information. This is elaborated in subsequent chapters. Briefly, an audit report should have this basic information:

- Scope of the audit (what processes were audited)
- Audit team members
- Time and location of the audit
- Individuals who were interviewed
- Narrative of activities that were assessed
- Good practices that were observed
- Risks that were perceived
- Problems and nonconformances that were found

Notwithstanding its legacy identity as the repository of a litany of "dings," an audit report can, and should, contain information that can be used to better understand how well an organization is operating. The report should provide a balanced representation of the status of the organization. The information contained in the report will be used to make decisions that affect the entire organization. It provides insight into the following:

- Processes and functions that are working well
- Perceptions of risk
- Detection of things that have gone wrong

From these insights should flow corrections, corrective actions, preventive actions, lean initiatives, benchmarking activities, and an array of improvement projects. All these actions require the allocation of resources. Think of it as the justification for expenditure of resources that should eventually reflect a positive ROI.

Factors and situations have been observed that need to be addressed. The objective evidence contained in the audit report provides the rationale and justification for the allocation of resources and the ensuing implemented actions. These, in turn, should result in improvements to the organization. If the information in the audit report is not complete and reflective of the processes audited, the auditee could easily arrive at an inaccurate impression as to the status of a process, the nature of a problem, or the risk that it carries. The ensuing actions may not appropriately address the given situation. Or, if the audit report is incomplete, it could result in no action being taken and the perpetuation of a negative trend.

The audit report encapsulates the fact-based foundation against which any ensuing action will be compared. When the time comes to assess the status and effectiveness of the actions taken, the audit report will become the yardstick against which the measurable improvement can be analyzed. If findings have been clearly articulated, the audit report will ultimately facilitate the determination of the extent of benefits (if any) that have been derived from the actions taken. If there have been findings of nonconformity or opportunities for improvement, top management will know that the actions taken have been based on good data—on objective evidence. The actions taken will relate directly to the audit findings.

Figure 1.4 shows another example of a process approach in action. The input is the finding. The executed action plan is composed of the activities. The results are the output, which can be assessed against the inputs.

Inputs	Activities	Outputs
Finding of nonconformity or opportunity for improvement	Implemented action plan as detailed in the following: • corrective action • preventive action • other improvement initiative	Assessment of the extent to which the plan has addressed the cause and the level to which improvement has been actualized

Figure 1.4 Process approach to nonconformity and improvement.

Finally, an audit report should start with a good plan. The output should reflect the rigorous deployment of that plan.

In order to have a credible audit report that creates value for the auditee, the auditor needs to plan the audit in anticipation of what will need to be reported. This means that the auditor's preparation should consider, as appropriate, the following:

- Standard being used as the audit criteria
- Evidence to be gathered
- Sample size of evidence
- Criticality of processes
- Number and types of individuals to be interviewed
- Alternate methods of assessing conformity to requirements
- Ancillary processes that need to be assessed

Proper planning should result in effective recording of the information that will become the audit evidence. This, in turn, supports the audit findings that will be used to substantiate the audit conclusion. It mitigates the risk of walking away from an audit with incomplete or inadequate information that unfortunately can lead to an incomplete or otherwise unhelpful audit report.

The essential fact to keep in mind is that the audit report needs to be informative and provide value. It's important to write a comprehensive audit report. The report doesn't have to be lengthy, but it should convey a balanced summary of the status of the organization audited. It should mention good practices that have been observed, risks that have been perceived, and problems that have been identified.

KEY TAKEAWAYS

- The audit report is the output of the audit process.
- The audit report should reflect the process approach model with inputs matched against outputs.
- The audit report should be comprehensive and provide a balanced representation of the status of the organization.
- The audit report should not be a simple litany of findings of nonconformity.
- The audit report should create value for the organization that has been audited.

Chapter 2

Purpose and Distribution of the Audit Report

As an auditor is writing a report, it's important for them to be attentive to the audience and the distribution. This attention is directly relevant to the purpose of the audit and the intended recipient(s). Who is going to get the report, and who is actually going to read it? The audience for audit reports varies. The network of individuals who will read an audit report is dependent on such factors as the purpose of the audit, the organizational culture, and company policies. Concerns to be considered while writing the report include confidentiality, proprietary information, ease of comprehension, and the need to address any findings. Throughout the preparation of the audit report, it's essential to bear in mind who will read the report and how they will receive the information it contains.

For internal audits, the distribution could be very limited. It might be that organizational policies direct that the quality manager reviews the report, makes it available to the relevant process owners, and at the same time generates and assigns any requests for corrective action. The quality manager might summarize key issues during a management review, but ultimately the report will never again see the light of day. In other organizations, the report might be available in a folder on a public drive that is accessible by almost everyone in the company. Variations might include having the complete audit reports as an addendum to the records of management review or simply having a log of the nonconformities and the opportunities for improvement.

PROTECTION OF CONFIDENTIAL INFORMATION

Proprietary information, regulatory requirements, intellectual property, and security protocols are all factors that require careful consideration. There must be balance between the inclusion of adequate objective evidence and the protection of sensitive materials. If the distribution involves individuals who do not have the requisite authorization, there may be a need to have two reports—the complete one with tightly managed distribution and a summary report with sensitive information redacted. I once observed an organization where the internal auditors (with the best of intentions) had been very thorough when including objective evidence. Unfortunately, some of the information they had included put them out of compliance with restrictions in their International Traffic in Arms Regulations (ITAR) agreement. A typical statement to avert unintended disclosure would be, "Some of the evidence used in the audit is protected by government regulations. If the need arises to get additional information, the reader is directed to make inquiry with authorized personnel."

This focus on protecting privileged information is also a prudent strategy when conducting a supplier audit. It is not uncommon to see copyrighted materials and patented processes. In fact, it is also possible to inadvertently observe competitors' information—or simply another company's intellectual property. Much of the risk of disclosure can be mitigated by using document numbers and other tracking identifiers without including individuals' names or titles that contain descriptors.

LANGUAGE AND TONE

Regardless of the distribution, tone and language are important. The tone of the report should reflect its purpose: to create opportunities for learning and improvement. With that stated objective, it's easier to get individuals engaged in the need to participate in a corrective action program.

Many years ago, I had a client who went through a certification audit. While the audit was a success and the company was recommended for certification, there was one minor nonconformance. The request for corrective action was routed to the process owner. He looked at the document sullenly. His body language spoke volumes. A brief conversation confirmed that he felt he was being blamed for the only blemish on an otherwise perfect audit. And he was being told to fix it! After listening to him, I explained, "This corrective action wasn't issued to you as a punishment. It was given to you because you understand these processes and machines better than anyone. We need your help."

His expression lifted, and he said, "All right. Then I will tell you the truth."

Implicit in the tone of the audit report is the auditor's overall judgment of the organization's commitment to its quality management system (QMS). Tone directly affects the recipient's attitude about the conclusions found in the report. This, in turn, predisposes the mindset with which ensuing actions will be conducted. In this case, the process owner had made a change to machine settings to improve efficiencies. It was a very good idea, but he hadn't communicated it adequately and didn't fully comprehend the downstream effects of the change. The finding simply made him feel like he had screwed up—and been humiliated as well. Without the follow-up conversation, he would have hidden his error and probably never made any other attempts to improve anything, which would have been unfortunate.

It is possible and imperative to write audit reports, particularly findings, with complete objectivity and respect. In an internal audit setting, the auditors and auditees are coworkers and associates. They may even eat lunch together. Acerbic comments or poorly supported hyperboles in audit findings can have a corrosive effect on a work environment.

Another audience to consider is top management. They may read the entire report, a brief summary, or only the audit findings. They must be able to comprehend quickly and clearly what the audit conclusion signifies. They should be able to easily garner salient information about the problems, the risks, and the benchmarks that have been illuminated by the audit. They should also be able to read a finding without leaping to the conclusion that one individual is responsible and needs to be reprimanded. Once again, objectivity diffuses the climate of guilt and ensuing punishment.

It's also important to remember that internal audits are essential inputs into the management review process. Some of the decisions that occur during strategic planning may be a direct result of audit findings. Therefore, what goes into the report matters.

The third audience is the individual or team assigned to work on the request for corrective action. In order to conduct a credible and thorough root cause analysis and be able to develop and implement an effective action plan, they need to fully comprehend the nature of the problem statement and its context. What happened? When did it happen? Is there potential for similar errors? How big is this issue? What caused it to happen? The objectivity and completeness provided in the statement of nonconformance are instrumental in launching the team into the corrective action process. Having a sense of the status of ancillary processes is also helpful.

For supplier audits, the tone might be more formal. However, the language makes the difference between a feeling of customer/supplier partnership or oppressed/oppressor confrontation. The outcome of the audit should result in both parties experiencing learning. In the case of supplier audit reports, the information might be used to decide if the company will be added to the approved supplier list. The language should internally convey a thoughtful conclusion as to the status of the organization's management system and its capacity to be a valuable addition to the supply chain. Externally, the report should express gratitude to the company for their responsiveness and willingness to make their system available for the audit. It should further state whether the company is being recommended for preferred status, if there are concerns that need to be addressed, or if they have unfortunately been deemed unqualified.

If there has been a degradation in overall performance, the summary might read, "Most processes remain well controlled; however, there is a concern with the receiving process that is raised and that has been articulated below in a request for corrective action." A more callous report might simply say that the overall score has dropped and the supplier has moved from approved to provisional—with no explanation except for abbreviated notes buried in the scoring checklist. This tone might appear cavalier to the reader and could damage an otherwise long-cultivated and beneficial supplier relationship.

A final tip to bear in mind: Always ensure commitment to objectivity. When stating a finding, auditors should endeavor to limit their use of adjectives and adverbs. They often tend to skew the significance of a finding beyond what objective evidence suggests or communicate an inappropriate note of disdain. Remember: It is better to use fewer words more precisely chosen.

KEY TAKEAWAYS

- The language and tone of the audit report should reflect consideration of the audience.
- The auditor should ensure that the information
    - is presented fairly and objectively;
    - refrains from suggestions of blame; and
    - contains language that is comprehensible to the reader.
- The language of the audit report should be clear and complete to facilitate any need for further action.
- The language and tone of an audit report should be respectful.
- Care must be taken to ensure the protection of sensitive and proprietary information.
- Audit reports can be summarized so that salient information can be made available to more people.

Chapter 3

Basic Information to Include

There's some basic information that needs to be included in every audit report. While it is very straightforward and seemingly mundane, this important information is essential.

Its most elemental function is to accurately record the who, what, when, and where of the audit. Each of these have their own relevance to the audit. This chapter deals with what must be included along with the rationale and potential pitfalls of omissions.

There is one salient caveat: Additional information may be required by the audit client (organization or person requesting an audit). This is commonly the case with third-party assessments, regulatory audits, or supplier audits. The additional information is dependent on the purpose of the audit and/or regulatory reporting requirements. With the exception of supplier audit reports (covered in Chapter 8), the reader is directed to reference the requisite information from the applicable audit client. These include certification bodies, accreditation bodies, governmental agencies, and regulatory authorities who each have their own unique additional audit report requirements.

This chapter describes the basic requirements that are common to all audit reports.

PURPOSE AND SCOPE

For internal audits, the purpose is straightforward. The purpose is to assess continued conformance to the organization's defined QMS processes. For a supplier audit, there might be several purposes:

- To assess the organization's QMS and evaluate its ability as a step toward prequalification as a preferred supplier
- To assess an organization in preparation for awarding a contract

- To conduct a periodic surveillance audit to determine continued conformance to requirements
- To follow up on corrective action requests
- To address a downturn in performance metrics
- To adapt to a major change in the company or to assess new processes or enhanced capabilities

The scope describes what will be audited. It simply means what processes or departments will be assessed. For example, an internal audit could include all applicable QMS processes in the production area on the third shift or the purchasing function for supplier qualification, processing of purchase orders, and incoming product acceptance—including material traceability. For a supplier audit, it could encompass all processes and QMS elements as defined in the scoring checklist or be limited to the shipping department and engineering office for the follow-up of two corrective actions from the most recent surveillance audit.

DATE OF THE AUDIT

When did the audit take place, and why is this information relevant? Recording the audit date serves multiple objectives.

First, the audit date as recorded can be compared against the established audit plan. This provides evidence that audits are being conducted in accordance with the audit program plan. Internal audits are an integral and important component of a management system. For most QMS schemes, they are a requirement. Regardless of certification status, without internal audits, a management system is incomplete.

Frequent lapses in the schedule may be indicative of an erosion in the organization's commitment to the internal auditing program. It might also reflect a resource problem. Are auditors being told by their supervisors that they can't take time away from their regular jobs to conduct an internal audit? This might suggest a devaluation of the process by some middle managers. Or it may simply be that there aren't enough auditors in the auditor pool, thus making it impossible to complete the audits within the allocated time.

Internal audits are not optional. If audits are seen as an intrusive and tedious drain on resources, individuals may be less apt to take them seriously. Audits that are conducted carelessly or as an imposed paper-shuffling mandate are less likely to yield any benefit for the organization. Third-party auditors, in particular, will note if there are apparent repetitive lapses and delays in the deployment of an audit program. Certification body (CB) auditors are required to verify that audits are being conducted in accordance with defined frequencies to ensure conformance to the organization's requirements as well as the requirements of any management system scheme to which it has been certified.

Noting the date provides a second benefit. If a process is being revised at the time of the audit, the date can help confirm which requirement was applicable at the time. For example, let's say a deviation has been approved to allow the use of obsolete inventory for a specified period of time. The audit date coincides with the allowable time period. It would be inappropriate to cite a finding if the material being used did not match the new bill of materials that engineering had recently released. If such a finding were raised, the auditee could justifiably object by pointing out that the material used during the audit was covered by the (signed and dated) deviation report.

In other instances, auditors are provided procedures prior to arriving on site, with the expectation that these are the current requirements. If there have been revisions, the only evidence that an individual was (or was not) adhering to current requirements would be matching the reported audit date against the document revision approval date.

For organizations that operate more than one shift, the time of the audit will also be important for all the same reasons noted in the previous paragraphs. One other key benefit is ensuring that there is uniformity of conformance and effectiveness across all shifts.

Finally, it's useful to record the duration of the audit. This helps top management get a clear picture of the resources that are being expended. How much time and money is the organization spending on the audit program? When compared against the effectiveness of actions taken as the result of audit findings, they can provide validation of the ROI. For example, analysis of data could provide the following information: "The resources expended to find problems and address risks yielded XXX dollars of savings." This holds true for both internal audits as well as supplier audits.

For certification audits, the time required to conduct the assessment is specified by the certification bodies' requirements. Condensing a three-day audit into a one-and-a-half-day visit is not allowable.

LOCATION OF THE AUDIT

For a small company, defining the location of the audit would be as simple as saying what departments were visited during the audit. However, a larger company may have an expansive campus with several buildings or multiple locations. In either of these case, it's important to record the location, especially if there are findings that the organization may wish to investigate further to see if they are localized or systemic. Conversely, observing good practices helps provide a benchmark so that they can later be applied in other parts of the organization. There's also the possibility that several specific activities are conducted at more than one site. Noting the locations that were audited will facilitate future planning to ensure the processes at any omitted sites are audited at a later date in the audit cycle.

STANDARD USED

The standards against which processes are audited include the organization's own documented procedures (in whatever media they are formatted), any applicable international standards or regulatory requirements, and the customer requirements. In order to ensure the accuracy of the audit report and any potential findings, it's important to plan what standards will be used. This will include not only the title or identifier of the document but also the revision level and/or approval date. Without this information, the validity of the audit report is jeopardized.

Let's take each one of these in turn:

1. Internal documentation

   These are the documents that define the requirements for the processes and product specifications. In today's organizations, especially those with a certified management system, the most common internal documents are procedures. They need to be controlled, accessible, and approved. Other examples include the following:

- Work instructions
- Manufacturing routers (also called travelers, process sheets, or route sheets)
- Requirements embedded in digital media, such as parameters developed through software that control a process or provide acceptance criteria
- Posted instructions and visual aids

2. Applicable management system or regulatory documents

These documents are developed by standards development organizations, such as the International Organization for Standardization (ISO), International Aerospace Quality Group (IAQG), and International Automotive Task Force (IATF), or by government agencies. The requirements may be voluntary, like ISO 9001, or mandated by state or federal statute, like the U.S. Food and Drug Administration (FDA).

Examples include the following:

- ISO 9001
- IATF 16949
- TL 9000
- ISO 14001
- 21 CFR Part 820

3. Customer specifications

Customer specifications may relate to the product being provided or could relate to additional requirements such as packaging, traceability of raw materials, allowable overages, licensing of staff, barcode labeling, provision of certificates of conformance, or third-party testing data. Since fulfillment of all applicable requirements falls within the scope of audits, it's important to cite them in the audit report.

Examples may include these elements:

- Customer drawings and specifications
- Unique booklets or files enumerating coded requirements, such as those noted earlier for traceability, certification, labeling, etc.
- Service contracts or statements of work (SOWs)

In each case, the lists of examples are not exhaustive. Depending on the industry, the kind of product or service being provided, or the customer's sector-specific concerns, there may be other documented requirements that are communicated in a variety of ways. What they all have in common is that they constitute standards against which an audit may be conducted.

It is reasonable to anticipate that there will be several standards associated with one process. For example, when auditing the purchasing function, the applicable standards might include the organization's purchasing procedure, the requirements embedded in digitized media (such as an ERP system), and the ISO 9001 (or comparable QMS scheme) requirements related to control of the externally provided products and services (supplier management and purchasing).

The audit report would then cite the following relevant documented information as the standards against which conformity was assessed:

- Procedure QP 46-003 Purchasing Rev C
- Work Instruction WI 46-005 Establishment and Maintenance of Inventory Levels Rev E
- ISO 9001:2015 8.4
- Automated fulfillment module in ERP system XXX
- Customer XXX document SUP-09 dated August 2024 for supplier requirements

Placement of the information about the standards that have been used can differ. Typically, the overarching standard is cited at the beginning of the document. For a certification audit, it would be the applicable management system standard—for example, ISO 9001:2015. For an internal audit, it would be the main procedure. In the example used earlier, it would be the purchasing procedure. All others can be listed later in the audit report, along with the records that describe the process and constitute the evidence, due to the fact that documents and records both fall into the category of "documented information" in current QMS models. This also makes sense because some documents eventually transition into records. An open purchase order is a document relating to requirements for purchased materials. A closed purchase order provides evidence that the product or service was correctly and adequately communicated to the supplier and that what was received conformed to the original requirement.

AUDIT TEAM MEMBERS

Every team has a lead auditor. If there's only one auditor, that person is the lead auditor. That individual has the ultimate responsibility for generating the audit report.

Other team members must be listed in the report, along with any technical experts who accompanied the team. Technical experts serve to provide very specific information that may exceed the knowledge of any of the auditors. These individuals are not generally used for internal audits unless the organization has some highly specialized processes.

There are other individuals who should be mentioned, although they are not specifically members of the audit team. They include observers and (for third-party audits) witness assessors.

Bearing in mind again that the audit process itself gets audited, having these names allows the auditor to confirm that none of the internal auditors audited their own work. When conducting surveillance audits, a third-party assessor may ask what tasks the auditors conduct in the company and compare them against the scope of the audits they have conducted.

PERSONS INTERVIEWED

Those who have answered questions should be identified. In some instances, to retain confidentiality, they may be denoted by their initials, first name, or job title. This is particularly important if the audit report is being electronically transmitted internationally. For example, the European Union (EU) has policies governing transmission of personal information. The General Data Protection Regulation (GDPR) specifies what can be shared. Refraining from using full names to identify the auditees facilitates conformance to those requirements.

There are a few advantages to recording the names of the auditees. These are generally individuals who either are the process owners or have responsibility for the activity.

It's not uncommon for people to try to be helpful and answer an auditor's question even if it's not part of their regular job. Sometimes the auditor finds out too late that she wasn't speaking to the right person. What an auditor may hear from a manager during a closing meeting is something like, "Francine doesn't take care of patient intake, so she wouldn't know where those forms are kept." As awkward as it is for the auditor, it's best to find out this type of information even late in the process so that corrections can be made and unwarranted findings of nonconformance can be removed.

The other advantage is that auditors can assess if activities are being conducted by individuals who have the authority, responsibility, and qualifications for the tasks they are performing.

Finally, recording the names of persons interviewed helps auditors provide objective evidence that they've fulfilled the requirements of the auditing process.

The chapters that follow cover the rest of the information that should be included in an audit report, covering these topics:

- Narrative of the audit
- Findings of nonconformity
- Opportunities for improvement and observations
- Verification and closeout of corrective actions

KEY TAKEAWAYS

- Good planning will help auditors prepare for the information that must be included in the audit report.
- Essential information for the audit report includes these elements:
    - o Purpose and scope
    - o Date
    - o Location
    - o Standards used for audit
    - o Audit team members
    - o Individuals interviewed
- Omission of any of this information impugns the integrity of the audit and may hamper the ability to take effective action.

Chapter 4
Narrative of the Audit

It's interesting to think of the audit report as a narrative. This is particularly true for second- and third-party audits. Frequently, for these types of audits, the auditee is relatively unknown. Even internal audits conducted at multisite organizations can include locations the auditor has never visited. The facility is an unexplored venue. The processes are sometimes unique or exhibit unanticipated innovation and creativity. The audit report then becomes the story of walking into this space and seeing how everything works. The auditor serves as the narrator who details the characters, location, and activities. In short, the report provides a narrative of the who, what, where, and when of the audit.

Replace the words "visitor" with "auditor," "characters" with "qualified staff," "location" with "infrastructure," and "activities" with "processes," and the parallel becomes evident. Writing the audit report should include that same narration with clear description of how well these factors interact and interrelate.

The details for each process can vary considerably depending on the organization. For a medical device company, the description of the infrastructure will be elaborate. It will provide evidence that processes are conducted in an environment that mitigates the possibility of contamination. In a healthcare setting, the qualification and training of individuals to interact with patients may require greater elaboration in the audit report. A manufacturing organization that has a simple assembly process will have less narrative than one with multiple and variable operations that may change with each customer order.

How, then, should an auditor present his report of the visit? What information should be included? What kind of details will provide adequate objective evidence that the auditee's process or entire management system conform to specified requirements—as defined in the referenced standards?

There is no particular formula for the order in which this information is presented. It's best to begin with a brief summary, followed by details of processes that have been observed.

It's important to say a few words about the audience. Auditors, despite the fact that they are bound by a code of ethics to both confidentiality and objectivity, should consider who will be reading the report. Members of top management are frequent readers. They can find within the pages of the report not only findings of nonconformity but also information about what processes are working well. Auditees may read the entire report or just the findings cited in their area. If the auditee is a supplier, the report may reflect whether or not they will be qualified as a new supplier or continue to enjoy preferred status. The tone and vocabulary of the report should accommodate the intended audience.

WHAT TO INCLUDE IN THE SUMMARY

The summary sets the stage and helps the reader understand what was observed and any circumstances that may have had an effect on the outcome of the audit. Items of note include the opening meeting, a summary of processes that were audited (along with some good points), positive observations, and the status of any corrective actions raised in the last audit.

Opening Meeting

In the opening meeting, describe some of the following items:

- Individuals who were in attendance.
- Exceptions or items of note that were discussed. Examples include the following:
  - The third shift doesn't run during the summer months.
  - A new ERP system has been installed recently.
  - The call center is now operating from a remote location.

- There were anomalies that affected the manner in which the audit was conducted, such as the following:
  - There was a power outage during the audit.
  - A key process owner was on vacation.
  - The auditors were required to participate in a 15-minute training before being granted access to the cleanroom.

None of these factors are findings. They are neutral facts. They will in some instances require the auditor to describe in the audit report what alternate approach or method was used to achieve the audit objective. For example, in the absence of a process owner, who is designated as responsible for decision-making, and how does the process continue to be well-controlled?

Over the years, there have been two instances in which I was the lead auditor at facilities where the company lost power. I requested that management find a room with sunlight and then rearranged the audit schedule so that individuals could bring records to the room. The purchasing time slot was swapped out for the electronic assembly operation. The interviews were conducted and the documentation assessed. The fact that this occurred needed to be noted in the audit report since the audit plan had changed and the location and manner of interviewing had been adjusted to accommodate the circumstances.

Therefore, things that didn't go according to plan but still resulted in a successful audit outcome should be noted. Notwithstanding the need to communicate anomalies that may have affected the audit, it's still possible to report this concisely without getting into too many details. And auditors should use their own judgment to decide which hiccups are really worthy of being reported. In some cases, it might be a simple as a comment that "the audit was conducted in accordance with the audit plan with slight modifications resulting from a temporary power outage."

Conversely, things that don't go according to plan and therefore adversely affected the audit should *definitely* be noted. Prior to showing up on-site, an auditor will have communicated the plan to the auditee. If, upon arrival, it is discovered that several processes are not running and multiple key personnel are absent, the auditor is required to report that part or all of the audit objectives could not be achieved.

Summary of Audited Processes

There should be a few brief sentences about what activities were audited (along with some good points). It doesn't have to be elaborate. Examples of simple statements are these:

- The audit process involved assessing three of the manufacturing operations, including the inspection and tests, qualification of operators, and review of applicable documentation.
- The audit's scope covered all the applicable requirements of the quality management system.
- All applicable processes and support activities for the third shift were assessed.
- The audit involved the storage and shipping processes at the company's remote warehouse.

Further detail will follow later in the audit where there are more elaborate descriptions of the process, the kinds of documents used, the availability of resources, and the appropriateness of the environment for the activities being carried out.

Positive Observations

The next thing to ensure is that positive observations are articulated. An audit isn't an attempt to amass a collection of bad incidents. Therefore, an audit report should also mention good points that were observed. Some examples include these comments:

- A newly developed software design program is facilitating communication and traceability between departments on new projects.
- The records provide evidence that operators who have had training on the ERP system have increased agility in accessing the files they need in the fabrication area.
- The nonconformance tracking system is better able to analyze data relating to scrap rates and late deliveries.

It's also nice to acknowledge accomplishments and success stories. People are so accustomed to only hearing about the dreaded "NCs" in the audit reports. They should also be able to read things like, "There has been significant improvement in the warehousing system" or, simply, "Staff should be acknowledged for a job well done."

Status of Previous Corrective Actions

The last thing to put into the summary is a brief statement about the status of corrective actions that emanated from the last visit. It's adequate to say that findings from the previous audit have been effectively addressed. This may actually tie in with the previously mentioned positive comments. For example, "In response to a corrective action originating from the last audit, the shipping department conducted a kaizen event that has significantly increased the accuracy of the picking process, resulting in a decrease in shipping errors." Unfortunately, an auditor may have to say that there has been no progress in addressing major issues that were previously raised.

PROCESSES AUDITED

The summary is followed by detailed descriptions of each process audited. In some instances, there will be only one process. The report should provide a basic description of the process, the documents that were assessed, the records that were reviewed, and the activities that were observed. Examples are provided in the following sections.

Call Center

Listened in on three customer calls while observing the data that the customer service representative input into the system. Two calls related to defective products, and one was for replacement parts.

Documents and evidence:

- Customer call records: 25-0769, 25-0781, and 25-0801
- Sample units used to troubleshoot the problem with the customer
- Multiple screens in Call Report software program

- Help guides to assist representatives
- Procedures QP 7-2-2 Rev J, QP10-2 Rev C, and work instruction WI7-2-4 Rev C

An opportunity for improvement is raised.

Fabrication Processes

Observed four different operations for the manufacture of five different parts. In each case, the operations are cut raw material, CNC machine parts, deburr, and box parts. The fabrication steps with specification requirements are described on the routing sheets. This is substantiated by the customer drawings. The other item is the setup sheet for the job. There were no jobs being set up at the time of the audit. For each operation, except boxing, a first-piece inspection is done. For the machining operations, sample plans establish the number of pieces inspected during in-process inspection. The verification of boxing is a visual check done by an operator from another area and occurs every two hours.

Documents and evidence:

- Job numbers: 23-6997, 24-7001, 24-7790, 24-6998, and 25-0003
- In-process inspection reports for job numbers listed prior
- Routing sheets for each job
- Drawing numbers: 23Z77712 Rev C, 89-NBV5506 Rev 9, 4478772 Rev B, and 23Z770044 Rev B
- Sampling plan WI 57734 Rev A
- Procedures: P4-05-9 Rev C, P4-05-7 Rev G, and P4-05-11 Rev D
- Calibration stickers on instruments used for inspection:
  - Dial caliper 1726
  - 1-inch micrometer 1709
  - Thread gauge 1822

  - Drop indicator 1622
  - 6-inch digital caliper 1719
- Inventory transfer tags for all jobs as listed prior

A finding is raised.

Design Process

There is only one current design project. It was initiated 18 months ago. The project is nearing completion. It was possible to review multiple records at each phase of the design project.

The design specifications are well developed. Acceptance criteria for each project objective are defined. This is substantiated by test methods. There have been four formal design reviews. Actions identified from each review are well tracked with results of actions taken. Records of tests conducted are all well organized. A trial run was recently completed, and the verifications are currently being done. The customer has signed off on design changes, artwork, and the installation manual. Validation will be conducted at the customer's location.

Documents and evidence:

- Procedure P-57832 Rev B
- Series of templates used to track each phase of the project
- Drawings, schematics, 3D CAD images, bills of materials, and raw material specifications for project XMM3-454
- Emails of discussions and decisions with customers about design changes
- Interim customer approval signature
- Gantt chart of project plan last updated September 30, 2025
- Records of design review for the dates of November 13, 2024, March 28, 2025, July 19, 2025, and September 29, 2025

There are no findings.

In each instance, the report has a brief description of what was observed and a listing of the documents and records that provide objective evidence to substantiate the audit conclusions.

CONFORMANCES AND NONCONFORMANCES

The final sentence in the report should state if the processes conform to requirements and if any findings have been raised. If the examples prior were all part of the same audit, it would be appropriate to conclude the audit report by stating, "The processes overall are well controlled. One nonconformance and one opportunity for improvement are raised."

The nonconformances can be included in the body of the audit report or documented on separate forms. Either method is acceptable. Writing up nonconformances is covered in Chapter 5.

In the case of a supplier audit or a certification assessment, there may be additional language about follow-up actions. Those relating to suppliers are dealt with in Chapter 8. For third-party assessments, auditors should refer to the rules of their certification body.

If the audit report is complete, it will reflect the process diagram discussed in Chapter 1, as shown in Figure 4.1.

Inputs	Activities	Outputs
• Audit scope • Audit criteria • Evidence of documents accumulated	• Interviewing • Observing • Assessing document • Reviewing records	• Evidence • Findings • Audit report

Inputs	Activities	Outputs
• Fabrication • Sampling plan WI 57734 Rev A	• Interviewed operators and machinists	• Routers for job numbers: 23-6997, 24-7001, 24-7790, 24-6998, 25-0003

• Procedures P4-05-9 Rev C, P4-05-7 Rev G, and P4-05-11 Rev D
• Production schedule
	• Observed in-process inspection
	• Assessed documents
	• Reviewed records of operation steps and inspections
	• Noted calibrated equipment used
	• Assessed documents used
	• Reviewed records
		• In-process inspection reports for jobs listed prior
		• Routing sheets for each job
		• Drawing numbers: 23Z77712 Rev C, 89-NBV5506 Rev 9, 4478772 Rev B, and 23Z770044 Rev B
		• Calibration stickers on instruments used for inspection:
		  o Dial caliper 1726
		  o 1-inch micrometer 1709
		  o Thread gauge 1822
		  o Drop indicator 1622
		  o 6-inch digital caliper 1719
		  o Inventory transfer tags for all jobs as listed prior
		• Findings (found later in report)
		• Audit report

Figure 4.1 Process diagram outlining findings of nonconformances.

KEY TAKEAWAYS

- The audit report provides a narrative of the who, what, where, and when of the audit.
- It's important to include information about the following:
    - The opening meeting
    - Any problems or deviations from the audit plan
    - Good things that were observed
    - Details about the processes audited, including these elements:
        - Documents that describe the process, including revisions
        - Documents that are used to describe specific requirements or plans
        - Inspection/acceptance methods and records
        - Activities observed
        - Records of completion of tasks or achievement of milestones
        - Approvals
    - Conformance to requirements
    - Any findings of nonconformity or opportunities for improvement

Chapter 5
Citing Nonconformities

Statements of nonconformities (NCs), also referred to as "findings," are frequent (although not inevitable) outcomes of audits. Sadly, they're often the only part of an audit report that gets any attention. The auditee rushes to the page where the NCs are listed, ignoring all the interesting text about the things that the organization has done well.

Prior to describing how to write up a statement of nonconformity, it's beneficial to review what the term actually means and how it relates to the audited requirements.

ISO 9000—Fundamentals and vocabulary provides the following definition of a nonconformity: "non-fulfillment of a requirement." ISO 9000 defines a requirement as a "need or expectation that is stated, generally implied or obligatory." Note 2 to the entry under this definition goes on to say, "A specified requirement is one that is stated, e.g. in documented information." The documented information can be in many forms, including, for example, an international standard, internal procedure, specification, customer requirement, or other kind of document that cites requirements. It also includes digitized requirements that are embedded in software programs, such as an ERP system or an automated manufacturing program.

Basically, a finding of conformity states that the organization should have produced a product or service to a specification or done something in a defined manner (usually documented), and it didn't.

There are four components to a nonconformance statement. We'll take each in turn. But first, it's important to cover some universally applicable guidelines.

GENERAL RULES FOR CITING NONCONFORMITIES

There are several general rules for documenting nonconformities.

- Be clear and concise. There's no need to be long-winded or verbose. However, it's essential to provide adequate information so the auditee understands what the problem is.

- Take the time to choose the precise words to describe what happened.
- Remember to tie the nonconformity to a requirement. If there is no documented requirement, all the auditor really has is something they don't like.
- Ensure that the text does not suggest blame. Whatever has gone wrong relates to the process—not the person.
- Be wary of presuming to know the root cause of a problem. This deprives the auditee of the opportunity to do an unbiased investigation to determine the cause.
- Refrain from suggesting solutions. Auditors should not usurp the auditee's right and ability to choose the appropriate action to be taken. Auditors should be careful not to articulate the finding in such a way as to take ownership of the solution. The auditee needs time and flexibility to do their own root cause analysis to determine the cause and to develop an appropriate and effective action plan. They understand their internal culture, processes, and capabilities better than the auditor does.

COMPONENTS OF A NONCONFORMITY STATEMENT

Bearing the aforementioned general rules in mind, let's break down the different parts of a statement of nonconformity.

1. What Is the Nonconformity?

This should be a clear, unbiased statement of fact that describes what's wrong. For example, "The inspections records provide evidence that material was accepted which exceeded the allowable tolerance range." Note that there is no accusatory tone or assignment of blame.

In the finding, there is no statement that a specific individual did something wrong. It would not be appropriate to say, "Michael passed material through that was out of spec." That statement presumes knowledge not in evidence that might become apparent during a root cause investigation. It could be discovered that there was a quality alert that Michael neglected to show the auditor indicating a temporary deviation to acceptance criteria. Or the instructions could have been verbal. The salient point is that the only thing that an auditor, in this case, can objectively state is that the inspection report indicates that material was accepted out of spec.

Another example of a nonconformity statement is, "There are no records to provide evidence that the report was reviewed by an authorized reviewer before being submitted to the client."

Again, there is no assignment of blame. Also, bear in mind that an auditor can't report what they have not observed. They can't say, "The report wasn't reviewed." The auditor doesn't know that the report wasn't reviewed—only that there's no record.

It's important to be starkly factual. The manner in which an auditor writes up a nonconformance will determine if the organization will chock it up to "operator error" and "retrain the operator" or if someone will ask why there was an error and look for the true cause, thereby preventing recurrence—or escalation to a more serious occurrence next time.

2. What Is the Requirement?

It is inappropriate—wrong, actually—to write up a finding of nonconformity that can't be tied to a requirement. Without a requirement, all an auditor has is something he doesn't like—and it's irrelevant to the audit report. The citing of the documented requirement justifies the finding. For the first example, the requirement is found in the customer specification. When citing a document, the auditor must be specific. The statement of requirement might read, "Drawing 7878993 Rev F calls out a tolerance of 9.75 inches +/− .005. The records indicate that the parts accepted measured 9.62 inches."

Looking at the second example related to reviewing a report, the report might state, "Procedure 8.2.4 Rev C specifies in section 7.7 that all reports must be reviewed by a second independent reviewer and that the reviewer must sign and date the last page of the report." Details like revision levels are also important. They sometimes shine a light on the fact that people don't have the most current information—steering the investigation away from the unfortunate and ubiquitous "operator error" root cause.

It is possible, in some cases, for there to be more than one documented requirement. It could be a combination of an international standard and a customer-specified instruction.

Having specific information as to the requirement provides several benefits. It demonstrates why there is a nonconformance. This, in turn, dispels any confrontations as to the legitimacy of the finding. And, finally, it helps launch the root cause analysis when it comes time to investigate the problem.

3. What Is the Evidence?

The third factor to be included is the evidence that substantiates the finding. For the first example, it would be Inspection Report #15549 from August 30; for the other, it would be Service Report #546 from September 15. There might be additional information, such as a lot number for materials or what shift was running the job. This both reinforces the justification for the finding and also facilitates the root cause analysis process.

4. What Is the Associated Risk?

This last component of a statement of nonconformity is optional. The finding is complete without it. But, the addition of a brief statement about the risk indicates that this is not an inconsequential matter. The auditor has not raised the finding trivially without understanding the level of risk it engenders. I call it the "so what?" factor. It's a brief statement of the reason this problem needs to be addressed. With both of the examples given, it would be appropriate to say that the risk is that customers will get defective product. Other concerns might include timely rework or possible regulatory action.

In the first instance, a statement about the risk might read, "The out-of-tolerance dimension would not be detected until the customer attempted to install the part into their instrumentation, which would result in costly delay for the customer and expensive time-consuming rework or replacement of the part."

For the second example, the risk statement might say, "Failure to have a signed review by an authorized individual could invalidate the report and result in a regulatory complaint and potential legal action."

Having a clear picture of the potential consequences of a recurrence of the situation can provide additional impetus for the organization to conduct a thorough root cause analysis and come up with an effective plan. It also demonstrates to top management that the ensuing corrective action has merit and is not simply a pencil-pushing exercise to appease an auditor.

The last thing to mention about thoughtfully worded nonconformity statements is that they provide a solid foundation in the event the finding is challenged. A poorly worded finding with vague reference to a requirement and incomplete evidence should be challenged. If an auditor finds themselves in a situation where an auditee wishes to contest a finding, having all this information clearly and completely documented in the audit report significantly increases the auditor's ability to defend the statement of nonconformity.

Figure 5.1 shows an example of how such a form could be filled out. (Sometimes the auditor's organization uses a separate form for reporting audit findings.)

Nonconformity Report

XYZ Company

Date: ______

Source of Request:
- ☐ Internal Audit
- x External Audit
- ☐ Customer Complaint
- ☐ Customer Observation
- ☐ Internal Observation
- ☐ Inspection
- ☐ Review of Analysis & Data
- ☐ Other

What is the nonconformity?	**What is the requirement?**	**What is the evidence?**
The inspection records provide evidence that material was accepted that exceeded the allowable tolerance range.	Drawing 7878993 Rev F calls out a tolerance of 9.75 inches +/- .005.	The records from Inspection Report #15549 indicate that the parts accepted measured 9.62 inches.

Risk: The out-of-tolerance dimension would not be detected until the customer attempted to install the part into their instrumentation, which would result in costly delay for the customer and expensive time-consuming rework or replacement of the part.

Initiated by: ______________________________

Figure 5.1 Example nonconformity report.

KEY TAKEAWAYS

- An audit finding should be an unbiased statement of facts.
- An audit finding of nonconformity should be clear, concise, and complete.
- Audit findings should not ascribe blame.
- Audit findings should not suggest root cause or offer solutions.
- Audit findings have four unique components:
  - Statement of what is wrong—the nonconformity
  - The requirement(s)—found in documents such as procedures and specifications
  - The evidence—the actual object or record that substantiates the finding
  - Risk—a statement describes the consequence of the nonconformance
- Well-written audit findings help the auditee by providing direction for their root cause analysis and eventual action plan.

Chapter 6

Opportunities for Improvement

Some things observed during the course of an audit do not meet the threshold criteria of a nonconformity. The process that has been observed does conform to the documented requirements. But the manner in which some aspect of the process is implemented suggests some affinity toward the essence of what is defined as the "nonfulfillment of a requirement." It cannot be denied that the process does technically fulfill the requirements. Nonetheless, something doesn't seem quite right or the manner of implementation engenders an unacceptable level of risk. These observations are often called "opportunities for improvement," frequently abbreviated as OFIs.

Let's step back for a moment and look at two of the auditing principles promulgated in ISO 19011. The first one relates to "Due professional care: the application of diligence and judgement in auditing." It states, "An important factor in carrying out their work with due professional care is having the ability to make reasoned judgements." An auditor exercises due deliberation when findings are raised. In citing both nonconformances and opportunities for improvement, they deliberate carefully, particularly in terms of criticality, level of risk, and prevalence. Findings should not be raised frivolously.

The second principle to consider is "Risk-based approach." The description of the principle states, "The risk-based approach should substantively influence the planning and implementation of the audit program and the planning and conducting and reporting of audits in order to ensure that audits are focused on matters that are significant for the audit client, and for achieving the audit program objectives."

The part of the second description that is salient to the discussion of OFIs deals with reporting the audit results. This principle of a risk-based approach to auditing has at its heart the expectation that findings that are raised will be converted into opportunities for improvement. Organizations often mitigate risks by taking actions that result in opportunities for improvement.

There may be a perception that a process is not adequately defined or controlled. This places it at heightened risk for an eventual nonconformance. There is a risk of the following:

- The customer will receive nonconforming product.
- There will be an internal failure that will result in loss of time, material, and money.
- The auditee will fall out of compliance with either standards (such as ISO 9001), industrial codes (such as NFPA building code), or regulatory requirements (such as FDA), putting it in jeopardy of losing its certification or license or experiencing some form of punitive action.

As auditors are assessing a process, they may perceive an activity that is implemented as documented but somehow engenders some level of risk. It may be that documentation hasn't been revised in several years and the text may be unsuitable for a new workforce that learns better through visual aids. It could be that an inspection sampling plan, while statistically valid, does not account for more stringent customer expectations.

These risks often arise because things change and processes, protocols, and documentation are not updated to reflect the changes. It's important to observe if some factor in the management system has changed. This is to be expected in any dynamic and well-implemented management system. Customer expectations evolve. New technologies emerge. Companies initiate innovative new processes. Nothing stands still. Changes need to be monitored.

Risks can arise as the result of an initiative that an organization implements to improve efficiencies. In that case, there may have been a failure to note if the action they've taken has increased the risk of things going awry in another department. Things have not gone wrong—yet.

It's appropriate for an auditor to make statements about perceptions of risk. These often identify a process that may not be controlled as well as it should be to prevent problems. OFIs do not say that something is wrong. They should clearly articulate the potential for something to go wrong. Examples might include, "It was observed that the router for fast turnaround jobs does not provide adequate instruction for certain steps that could result in errors and defective products needing to be reworked" and "Nonconforming material waiting to be scrapped is stored in close proximity to customer property. Even though the material is tagged, it resembles the customer's product, and there is a risk of the customer property being accidentally discarded." Again, in each example, there is no nonconformance. There is, however, the risk that something could easily happen that could result in a problem or nonconformance.

The important thing is for the auditor to ensure that the opportunity is not articulated as a recommendation. Auditors are not consultants. Making recommendations robs the organization of the learning process implicit in conducting their own investigation of the situation. It does not allow them the benefit of research and discovery. It stifles their ability to involve individuals in the investigation as well as the development of a viable plan of action. This eventually leaves their system incomplete because they aren't utilizing the established processes intended to help them learn, innovate, and improve. The auditor ends up owning (and inadvertently dictating) a solution when the onus is on the auditee to assess the perception of risk and determine what action (if any) is appropriate.

However, there are instances in which the audit client (the entity that has requested the audit) may have requested that recommendations be included in the report. For internal audits, top management may have requested that the internal auditor include recommendations due to their technical expertise or specific knowledge of a process. The other instance would be a supplier audit, wherein the auditor has been asked to include recommendations so that the supplier is better able to comply with the organization's needs, thereby mitigating the risk of losing a valued source for an important product or service.

However, it is important to reiterate that auditors should avoid consultancy.

Offering OFIs isn't a bad idea. In fact, they can be very helpful, provided they're worded judiciously. They should communicate a perceived risk. The message conveyed is that something appears to be on shaky ground, and it's worth taking the time to have a meaningful discussion and explore the need for action. A typical example might be, "It is unclear if the work instructions relating to the field installation are adequate for the anticipated new hires." There is no nonconformance.

Everyone currently using the instructions has been installing the product for five or more years. But the documents don't have many illustrations and are in a very formal text format. According to the records of management review, the organization is planning a major expansion and is going to be hiring twelve entry-level field technicians. The risk the auditor has communicated is that the work instructions may be okay for now, but they may not be effective in three months. These concerns point to specific requirements dealing with planning, allocation of resources, training needs, and effectiveness of documentation. In this way, they are anchored to the requirements of the standard against which the audit is conducted. They aren't haphazard opinions based on an auditor's whim.

OFIs don't tell auditees what to do; they give them something to think about. They give subtle direction. The ensuing decision and action belong to the auditee.

Auditors should be catalysts for change. But they should not be the change agents. So, when formulating the audit report, it's important to ensure that the auditee understands the distinction between a nonconformity (a nonfulfillment of a requirement that must be addressed) and an opportunity for improvement (a perception of risk that should be considered).

Figure 6.1 offers a quick comparison between an observation of nonconformity and an opportunity for improvement.

Nonconformity		Opportunity for improvement	
Part number XJ34-778 was located on an unmarked bench and had no tag or marking indicating that it was nonconforming. The parts had not been logged in the NC log describing authorization or description.	ISO 9001 8.7 requirements relating to the following: – Segregation – Documented information – Authorization	Part number XJ34-779-CUS was located on a marked bench, properly tagged and logged. However, the adjacent bench is used to store similar parts that are customer owned. Due to space constraints, there is a risk of comingling parts, resulting in the customer getting nonconforming product back.	ISO 9001 8.5.3 requirements relating to identification and protection of customer property.

Figure 6.1 Comparison between NCs and OFIs.

The final thing to remember is that the level of perceived risk should be significant. There are dozens of things we can all imagine could go wrong. The questions are how likely they are to occur and what the effect would be. If the risk is preposterously remote or the effect would be inconsequential, an OFI is a waste of time. Citing an OFI should provide the opportunity to mitigate actual significant risks and for the organization to experience learning and improvement.

KEY TAKEAWAYS

- OFIs articulate observations of risk.
- OFIs observe that the following may occur:
    - Customers might get nonconforming product.
    - There may be an internal failure resulting in lost time or money.
    - There is a risk of falling out of conformance with mandated requirements resulting in punitive (or other) action.
- OFIs are not nonconformities.
- OFIs should not be worded as recommendations.
- OFIs should articulate significant observations of risk.
- OFIs should not be trivial.
- OFIs provide the auditee with opportunities for learning and to mitigate risks.

Chapter 7

Verification and Closeout of Corrective Actions

There is a cyclical aspect to almost all audits. The whole concept of an audit program implies that audits are never singular or standalone events. Therefore, it logically follows that, consistent with the cyclical nature of audits, the findings from a previous audit create an input into the next successive audit.

Since follow-up is an intrinsic component of an audit, it is reasonable and appropriate to include narrative about the assessment of the actions taken to address the prior nonconformances within the audit report. This could include a conclusion as to the effectiveness of the action taken and the decision to close out the finding. Conversely, it might instead require a statement that no action was taken or that the action taken was not effective in addressing the root cause of the initial finding. In that case, the decision as to what further action is to be taken resides with the audit client. This is generally outside the scope of the audit report, unless policy dictates that specific action is communicated in the report.

VERIFICATION

Verification is generally a requirement for third-party audits. During the course of a surveillance audit, the certification body auditor will incorporate time into the audit plan to assess the results of corrective actions taken on findings from previous assessments.

Verification of corrective actions has two components: verification of *implementation* and verification of *effectiveness.* In short, did they implement action in accordance with their corrective action plan? And did what they implement work? That is, did it address the root cause and prevent recurrence of the problem or defect?

For internal audits or supplier audits, verification is sometimes conducted as a separate event. This is particularly the case if the finding has been graded as major or has been determined to carry an exceptional level of risk.

However, in most instances, it's more efficient and more effective to include it as part of the next audit. The auditor can observe more completely how the corrective action has been integrated into the processes of the quality management system. This provides for a more thorough assessment of the manner in which the action has been implemented and an opportunity to note any additional benefits that have been derived.

It is also an effective and efficient practice to seize the opportunity of an internal audit to assess a corrective action that originated from a customer complaint or the detection of a problem or defect. In this way, the corrective action can be observed as part of the entire process rather than as some uniquely isolated event. If the processes all work well together, the integrity and effectiveness of the interrelations of the system can be observed.

There are three ways that the results of verification of corrective actions can be integrated into the audit report: as part of the narrative, as a unique section, or on a separate form.

Verification Incorporated Into the Narrative About the Process

The verification is incorporated into the narrative relating to the assessment of a process. A nonconformity statement in the previous chapter dealt with inspection reports. As the process is being audited, the corrective action can also be assessed. The audit report would include a unique paragraph dealing with the issue. For example, a statement might read, "The manner in which in-process inspection reports are managed has significantly improved. As a result of the corrective action taken from a previous audit finding, a step has been added to the process that requires a secondary review of the inspection sheet before a job can proceed to the next operation. Observed the in-process inspection sheets for jobs 23-898, 23-902, 23-916, and 23-925. Additionally, reviewed the data from the monitoring of in-process defects and customer returns for this and comparable issues. The evidence shows a substantial increase in the metrics on first-past yield and zero returns from customers for this problem. Also reviewed revised procedure QP4-8-1 and training records for this issue. The corrective action has been implemented and verified as effective."

Verification as a Unique Section in the Audit Report

In this instance, the organization may have an audit form with a unique section for recording the verification of corrective actions. The form would have a header with the original date of the finding, the corrective action tracking number assigned, and some abbreviated text about the original finding. This would be followed by text that would be similar or identical to the previous example.

Verification Recorded on a Separate Form That Is Appended to the Audit Report

In this third variation, the organization has either a unique form for corrective action verifications or a series of fields in their internal corrective action form for recording all the information needed to verify and close out the finding.

Figure 7.1 is an example of a form to record the verification of a corrective action.

Corrective action number:	Date initiated:
Nature of finding	
Verification evidence	
Documents reviewed	Evidence assessed
Summary of verification and conclusion	

Closeout approval: ______________________ Date: __________

Figure 7.1 Blank CAR verification form.

Filled out, it might look like this (Figure 7.2):

<table>
<tr><td>Corrective action number CA 25-02</td><td>Date initiated 3-15-25</td></tr>
<tr><td colspan="2">Nature of finding
The inspections records provide evidence that material was accepted that exceeded the allowable tolerance range.</td></tr>
<tr><td colspan="2">Verification evidence</td></tr>
<tr><td>Documents reviewed</td><td>Evidence assessed</td></tr>
<tr><td>QP4-8-1 Rev C</td><td>☐ In-process inspection sheets for jobs 23-898, 23-902, 23-916, and 23-925
☐ First-pass yield metrics for previous six months
☐ Customer returns log
☐ Training records for QP4-8-1 Rev C</td></tr>
<tr><td colspan="2">Summary of verification and conclusion</td></tr>
<tr><td colspan="2">The evidence shows a substantial increase in the metrics of first-past yield and zero returns from customers for this problem. It has had the additional benefit of positively affecting overall on-time delivery. QP4-8-1 has been revised, and training was completed in late July over the course of three sessions, accommodating all shifts. The corrective action has been implemented and verified as effective.</td></tr>
</table>

Closeout approval: *D. Brownwell* Date: *10/8/25*

Figure 7.2 Example completed CAR verification form.

This provides another example of the process approach in action. The input is the finding. The review of the implemented action plan involves the activities. The results are the outputs, which can be assessed against the inputs (see Figure 7.2).

Inputs	Activities	Outputs
• Corrective action request • Root cause analysis • Corrective action plan	• Interview process owners or staff • Review revised documentation • Review records related to both original finding and improved process or product • Review statistical data • Review records of training or any actions taken	• Determination that the action has been implemented • Conclusion that the action has prevented recurrence of the problem • Statement of any additional improvements or benefits observed

Figure 7.2 The process approach.

This provides the opportunity to bring the cyclical aspect of auditing full circle. Factors and situations have been observed that need to be addressed. The next audit provides an opportunity to assess actions and determine outcomes, preventing the perpetuation of negative trends that result from unique, disjointed citations of nonconformities.

KEY TAKEAWAYS

- Follow-up of corrective actions is an appropriate audit activity.
- Results of the assessment of corrective actions taken should be included in the audit report.
- Integrating the assessment of corrective actions and their results into the narrative of the audit report serves to support the conclusion that the action has been effectively woven into the quality management system.
- At times, it may be necessary to report that appropriate or effective action has not been taken.

  - It is the decision of the audit client what further action is needed.
- The records of implementation and verification should include these elements:
  - Statement of the effect of the implementation
  - Evidence in the form of the following:
    - Documents reviewed
    - Records assessed and practices observed
  - Any other benefits that have been observed
- There should be a conclusion and closeout of the finding of nonconformity.

Chapter 8
Supplier Audit Reports

Supplier audit reports carry all the information that has been mentioned in the previous chapters. There's the basic information about date, location, scope, documents, evidence, and so on, but there is a slightly different spin on the purpose and the anticipated outcome.

Supplier audits are generally conducted for six reasons:

1. Initial qualification or prequalification
2. To award a contract
3. Periodic surveillance
4. In response to a corrective action request
5. Due to a downturn in performance metrics
6. Due to a major change in the company

This array of disparate reasons for conducting audits results in some variations from the audit reports described in previous chapters. Out of this list of reasons, there are three that frequently require a unique format: audits for determining initial qualification, awarding contracts, and periodic surveillance.

SUPPLIER AUDIT REPORT FORMAT

In all cases, suppliers or potential suppliers are actually—or indirectly—vying with their competitors for business. This compels the audit process and the resulting reports to reflect a particular level of parity in the assessment process. All will need to be evaluated based on the same criteria.

Hence, the method used most frequently is the scoring checklist. And, unlike other audits, the checklist is also structured to ultimately encompass the audit report. Again, that doesn't obviate the need to include most of the information described in previous chapters. The header will contain the boilerplate information as noted earlier. The summary narrative is typically inserted into a field at the beginning of the report, although the choice of its placement can be driven by organizational preference or software constraints.

In many instances, all the data (including evidence and commentary) is input into either a spreadsheet format or a software program that accommodates the need to calculate the scores and generate the recommendation. These programs may additionally have links to a module for generating requests for corrective action.

The manner in which documents and evidence get recorded is markedly different. Generally, both documents and evidence are inserted into the field in the column and row that corresponds to a specific requirement. For example, requirements related to purchasing would be listed in the left-hand column. Subsequent columns would provide for grading and scoring, with the final column allowing for the insertion of text related to the corresponding procedure number and title, the supplier qualification records assessed, the purchase orders reviewed, any other supporting information, and auditor comment.

The numbered columns typically have headings such as these:

- Not implemented
- Minimal compliance
- Meets expectations in a systemic manner
- Exceeds expectations or regularly initiates improvements

The titles of these headings vary greatly with fewer or greater gradations. There are a multitude of titles for the column headings, but they all have the same basic methodology and intent that allude to the level of maturity for the process or element. In some cases, the numbering codes are strictly applied, while in others, the auditors have some latitude and can utilize their own judgment in scoring.

SCORING CRITERIA

The auditor also must ensure that the scoring is correct. It's important that a finding related to document control (for example) isn't inadvertently duplicated in another section, such as in the area where the document was being used for a particular process. If the scoring checklist was constructed to mirror ISO 9001 clauses, it is conceivable that an auditor would score low marks under 7.5 for QMS documented information. Subsequently, another member of the audit team could again raise a finding, for example, under 8.2.1 for control of documents related to a customer's specification. The result would be a double strike and an inappropriately lower score. This can engender serious consequences for the supplier if their continued status is dependent on a qualifying score—or if the awarding of a contract hangs in the balance.

Figure 8.1 offers an example of a completed scoring checklist. A sample form is found in the appendix.

However, it's first important to discuss the aspects of the supplier audit report that relate to the outcome of the audit, including summary, recommendations, and requests for further action. Notwithstanding the previous discussion of scoring checklists, there are many instances in which the typical audit report format described in previous chapters serves adequately to fulfill the audit objective. In that case, the summary, description of processes, listing of evidence, and citation of findings would be input into a less rigid audit report format, without the scoring element.

The summary should reflect the purpose of the audit. If it is to qualify a supplier, that decision should be noted in the audit report. Alternately, if the report must be reviewed by other members of the procurement team before a decision is made, that fact should be noted in the report. There should also be an anticipated date for communicating the result of the deliberation and the ultimate decision.

SURVEILLANCE REPORTS

If the purpose of the audit is to conduct a periodic surveillance or close out the corrective actions from a previous visit, the summary should offer a conclusion as to the continued conformance to requirements or the effectiveness of corrective actions that have been reviewed. It may also include recommendations or requests for additional action. This serves to optimize the value both parties experience from the supplier/customer relationship.

Typical statements might include the following:

> "The evidence presented demonstrates that the corrective action has been successful in addressing the problem and there has been significant improvement in the process."
>
> "The result of the surveillance audit reveals that the organization continues to maintain an effective management system."
>
> "The supplier will be granted provisional status contingent upon the completion of two corrective action requests found in this audit report."
>
> "Although the supplier conforms to all requirements of their QMS, they do not include a process for pest control. While this lapse will not deter prequalification, implementation of a pest control program will be required for full qualification and/or preferred supplier status."
>
> "The company has achieved an overall score of 96 out of 100. It is recommended for continued qualification as a supplier."

Company: Zoonomics Tools, Inc. Address: Palisade Ave. Canterbury, UK Date of audit: 8 October, 2025	Summary: The organization continues to maintain an effective QMS. The corrective action from the previous visit has been addressed through a kaizen event involving a cross-functional team of 12. This resulted in improvements to processes in two other areas.
Leader auditor: Angelina Souza	Score: 95 out of 104
Recommendation: Continued status as preferred supplier	

Figure 8.1 Example completed supplier scoring checklist.

Item number	Requirement	(1) Not implemented	(2) Minimal compliance	(3) Meets expectations in a systemic manner	(4) Exceeds expectations or regularly initiates improvements	Evidence and comments
1	Organizational chart				X	Current; updated September 10, 2025.
2	Management review conducted at defined frequencies				X	Two conducted annually—March and September. Good supporting data and risk analysis.
3	Quality objectives defined and monitored				X	Monitored and posted monthly.
4	Document control system implemented and maintained				X	Procedures all current. No changes required. Work instructions revised after kaizen event.

Item number	Requirement	(1) Not implemented	(2) Minimal compliance	(3) Meets expectations in a systemic manner	(4) Exceeds expectations or regularly initiates improvements	Evidence and comments
5	Customer documentation controlled and protected				X	Separate secure vault on server for customer drawings and specifications.
6	Engineering change process control				X	Reviewed ECOs 783, 787, 788, 789.
7	Software programs and electronic media controlled and backed up				X	Backups and accessibility consistently implemented and maintained.
8	Records retained, accessible, protected			X		All records found in accordance with defined requirements.
9	Competence requirements defined; staffing needs to be reviewed		X			Job descriptions haven't been revised recently, despite new production process.
10	Training conducted and records maintained				X	Training needs have been assessed and more training planned.
11	Infrastructure maintained and resource requirements addressed			X		Maintenance is adequate.

Item number	Requirement	(1) Not implemented	(2) Minimal compliance	(3) Meets expectations in a systemic manner	(4) Exceeds expectations or regularly initiates improvements	Evidence and comments
12	Customer order processing and communication				X	Customer service diligent in reviewing requirements. Looked at three orders and acknowledgments to customer.
13	Review of additional requirements			X		None noted.
14	Supplier qualification				X	All supplier audits up to date. ASL is current.
15	Purchasing records				X	Purchase orders contain all requisite info. Reviewed POs 15634, 15644, 15701.
16	Incoming inspection/ product acceptance			X		Receiving records all complete, except one material certification not in folder.
17	Production operations and/or service processes (list separately and score)				X	Reviewed jobs 25-03, 25-04, 25-17, and 25-22 for processes as noted following.
18	– Material cutting and bending				X	

Item number	Requirement	(1) Not implemented	(2) Minimal compliance	(3) Meets expectations in a systemic manner	(4) Exceeds expectations or regularly initiates improvements	Evidence and comments
19	– CNC machining				X	
20	– Deburring and polishing				X	
21	– N/A					
22	Calibration program		X			Calibration program stable. Reviewed multiple tools. Two verniers had expired stickers. However, recent calibration records are available.
23	In-process and final inspection				X	See prior for in-process jobs. Also, reviewed final inspection records for two additional jobs.
24	Shipping			X		Uncomplicated, straightforward process.
25	Inventory (ID, storage, traceability)				X	All well controlled.
26	Control of nonconforming product (segregation, identification, disposition)				X	All well controlled, segregated, and ID'd.

Item number	Requirement	(1) Not implemented	(2) Minimal compliance	(3) Meets expectations in a systemic manner	(4) Exceeds expectations or regularly initiates improvements	Evidence and comments
27	Corrective actions (root cause analysis, planning, implementation, and closeout)				X	Three recent corrective actions. Two have been verified for effectiveness; one is still open. CAs 199, 202, 213.

KEY TAKEAWAYS

- Supplier audits have some variations but contain the same basic information:
  - Purpose
  - Scope
  - Audit team
  - Date and location
  - Standard(s) used
  - Individuals interviewed
- Supplier audits are conducted for several purposes:
  - Initial qualification or prequalification
  - To award a contract
  - Periodic surveillance
  - In response to a corrective action request
  - Due to a downturn in performance metrics
  - Due to a major change in the company
- Scoring checklists provide a different format for recording the outcome of the audit.
- Auditors must be diligent in ensuring that scoring checklists are used consistently across all suppliers audited.
- Supplier audit reports may include recommendations or requests for action to optimize the value both parties experience from the supplier/customer relationship.

Chapter 9
Conclusion

There are many considerations that go into writing an effective audit report. The attention that is given to these considerations is directly relevant to the value the report will have for the organization. Good planning prior to the audit helps ensure that pertinent information is gathered and is thoroughly and honestly reported.

The audit report is the output of the audit process. It should be comprehensive and provide a balanced representation of the status of the organization—not simply an itemization of findings of nonconformity. Findings should not be raised trivially. They should be based on objective evidence and cited against documented requirements. The report should create value for the organization that has been audited, giving it opportunities to both learn and improve.

The language and tone of the audit report should take into account the audience. It should be clear, comprehensive, and complete. The language and tone should be respectful, and the report should be written in such a way as to be comprehensible to the reader. Information should be presented fairly and objectively, with no insinuation of blame. It should facilitate any further action that is needed and be a vehicle for improvement.

Basic information intrinsic to any audit report includes the purpose, scope, date, location, applicable standards, team members, and individuals interviewed. Omission of any of this information impugns the integrity of the audit.

Care must be taken to ensure the protection of sensitive and proprietary information.

The manner in which the audit report is written should reflect the process approach model with inputs matched against outputs. The audit report provides a narrative of the who, what, where, and when of the audit. It should mirror the processes that have been audited with information relating to the documents reviewed, activities observed, and records of verification and completion.

The auditor's observations should include good things that were observed, risks that were perceived, and problems that were identified. This provides the auditee with necessary information to make decisions going forward. These may relate to benchmarking good practices, taking actions to mitigate risks, and corrective actions for detected nonconformances.

In some instances, the audit report will include a request for corrective action, while in other cases, the corrective action requests are issued by the audit program manager after the audit report has been reviewed. In either instance, the audit report from any subsequent audit should include evidence of verification of the effectiveness of the corrective actions that were implemented, along with any other improvements that have resulted.

Supplier audits often make use of a scoring checklist. This checklist constitutes the format for the audit report. While the process of auditing is similar, the scoring method drives the manner in which findings are cited and how the overall conclusion is presented to the auditee.

The audit report serves the ultimate purpose of ensuring that the organization reaps the benefits of the process: learning, problem solving, risk mitigation, benchmarking, and improvement.

Appendix A
Blank Nonconformance Report

Nonconformity Report

XYZ Company

Date: ______

Source of Request:
- ☐ Internal Audit
- X External Audit
- ☐ Customer Complaint
- ☐ Customer Observation
- ☐ Internal Observation
- ☐ Inspection
- ☐ Review of Analysis & Data
- ☐ Other

What is the nonconformity?	**What is the requirement?**	**What is the evidence?**

Risk:

Initiated by: ______________________________

Appendix B
Sample CAR Verification Form

<table>
<tr><td>Corrective action number:</td><td>Date initiated:</td></tr>
<tr><td colspan="2">Nature of finding:</td></tr>
<tr><td colspan="2">Verification evidence</td></tr>
<tr><td>Documents reviewed</td><td>Evidence assessed</td></tr>
<tr><td></td><td></td></tr>
<tr><td colspan="2">Summary of verification and conclusion</td></tr>
<tr><td colspan="2"></td></tr>
</table>

Closeout approval: ______________________ Date: _________

Appendix C

Blank Supplier Scoring Checklist Audit Report

Company: ______________ Address: __________ Date of audit: _________	Summary:
Leader auditor:	Score:
Recommendation:	

Item number	Requirement	(1) Not implemented	(2) Minimal compliance	(3) Meets expectations in a systemic manner	(4) Exceeds expectations or regularly initiates improvements	Evidence and comments
1	Organizational chart					
2	Management review conducted at defined frequencies					
3	Quality objectives defined and monitored					

Item number	Requirement	(1) Not implemented	(2) Minimal compliance	(3) Meets expectations in a systemic manner	(4) Exceeds expectations or regularly initiates improvements	Evidence and comments
4	Document control system implemented and maintained					
5	Customer documentation controlled and protected					
6	Engineering change process control					
7	Software programs and electronic media controlled and backed up					
8	Records retained, accessible, protected					
9	Competence requirements defined; staffing needs reviewed					

Item number	Requirement	(1) Not implemented	(2) Minimal compliance	(3) Meets expectations in a systemic manner	(4) Exceeds expectations or regularly initiates improvements	Evidence and comments
10	Training conducted and records maintained					
11	Infrastructure maintained and resource requirements addressed					
12	Customer order processing and communication					
13	Review of additional requirements					
14	Supplier qualification					
15	Purchasing records					
16	Incoming inspection/ product acceptance					

Item number	Requirement	(1) Not implemented	(2) Minimal compliance	(3) Meets expectations in a systemic manner	(4) Exceeds expectations or regularly initiates improvements	Evidence and comments
17	Production operations and/or service processes. List separately and score.					
18	– Material cutting and bending					
19	– CNC machining					
20	– Deburring and polishing					
21	– N/A					
22	Calibration program					
23	In-process and final inspection					
24	Shipping					

Item number	Requirement	(1) Not implemented	(2) Minimal compliance	(3) Meets expectations in a systemic manner	(4) Exceeds expectations or regularly initiates improvements	Evidence and comments
25	Inventory (ID, storage, traceability)					
26	Control of nonconforming product (segregation, identification, disposition)					
27	Corrective actions (root cause analysis, planning, implementation, and closeout)					

www.ingramcontent.com/pod-product-compliance
Lightning Source LLC
La Vergne TN
LVHW022021200926
848481LV00007B/2467
* 9 7 8 1 6 3 6 9 4 2 3 5 3 *